diving and digging for
GOLD
Mary Hill

NATUREGRAPH PUBLISHERS

Library of Congress Cataloging in Publication Data

Hill, Mary, 1923—
 Diving and Digging for Gold.

"A handbook on finding, processing, and selling gold with historical information on gold mining."
 Previous ed. issued in 1960 by Pages of History, Sausalito, Calif.
 1. Gold mines and mining. 2. Hydraulic mining. [1. Gold mines and mining. I. Pages of History, Sausalito, Calif. Diving and Digging for Gold. II. Title.
TN421.H54   622'.342   73—22389
ISBN 0-87961-005-0

Copyright © 1960 by Pages of History
Copyright © 1974 by Mary Hill

Naturegraph Publishers, Inc.
3543 Indian Creek Road
Happy Camp, CA 96039
U.S.A.

Books for a better world

ACKNOWLEDGMENTS

To the many who have helped in the preparation of this book our special thanks. We wish to credit the following illustrations: The old woodcuts on pages 8 and 27 are from the sixteenth-century book *De Re Metallica*, by Georgius Agricola. On pages 15 and 24 are copies made of sketches drawn in the early 1840's by G.M. Waseurtz af Sandels. The originals are in the collection of the Society of California Pioneers in San Francisco. The old prospector on page 17 is from a Frederic Remington painting. The drawings on the lower halves of 26 and 27 are from Charles Nordhoff's book *California, A Book for Travellers and Settlers*, published in 1873. The dry washer on page 26 is modified from a photo in *Arizona Gold Placers and Placering*, published in the Bulletin of the University of Arizona, Vol. 8, No. 2. The diver on page 30 was sketched from a photograph in the Fifteenth Report of the State Mineralogist of the California State Mining Bureau, 1916. The equipment shown on pages 32 and 33 was drawn from a catalog of U.S. Divers, obtained from the Surf Shop in San Francisco. All other drawings are by Elinor H. Rhodes.

Our thanks to Mr. John Gow of Western Knapp Engineering Company, and to Mr. John Jan of Western Machinery Company, for the design of the retort shown on page 35.

"To Betty"

TABLE OF CONTENTS

Placers—What They Are 8

What to Take 16

Where to Go 18

Tips from Old Timers 22

Tools 24

Amalgams and Retorts 34

Life and Death of a Mine 37

Where to Sell Gold 42

Where to Get More Information 46

A Currier and Ives reproduction obtained through the courtesy of the Library of Congress.

PLACERS .... what they are

Prospecting with a divining wand, 1556

Placer mining (rhymes with plaster, not pacer) is an art as old as civilization.

No one knows for sure what the word *placer* originally meant. During the California Gold Rush, when Mexicans, Chileans, and other Spanish-speaking peoples headed for gold with the rest, they used their word *placer* to refer to gold deposits in sand bars and banks. Because there was no other good word to use, all miners accepted it. What was originally meant in Spanish is doubtful: some say that it is a local, perhaps slang word for *sandbank;* Spanish dictionaries define it—in this technical sense—as *"the place near a river where gold dust is found."*

But in most dictionaries, the first translation given for *placer* is *pleasure*. It is in this sense we hope you will want to define placer mining.

Gold is found in placer deposits because nature has sorted heavier fragments from lighter ones. In water a fragment of heavy gold will fall much faster than a rock fragment of the same size and shape.

The gold that is found in streams—placer gold—has been accumulated by dropping through water. Once it is worn away from its parent rock vein, it is borne by creeks and rivers toward the sea. As it travels, it gradually sinks toward the bottom of the stream to find a resting

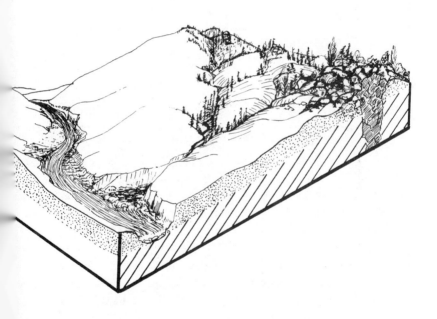

Block diagram showing broken outcrop of gold-bearing vein. The enriched soil is a "seam digging". The gold is washed into the streamlets to be moved to the river, where eventually it may become a placer deposit.

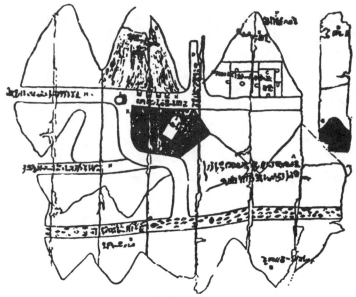

Papyrus map made about 1300 B.C. The hieroglyphics mean "mountains in which gold is washed", "houses of the gold washing settlement", and "mountains of gold".

place at the lowest point it can reach. Obviously, creeks and rivers are not still water, and their turbulence, their power, their swiftness, and the nature of their course and bed all help determine the place the gold will rest.

Nature releases the gold in her hard-rock storehouses in a number of ways. She splits the rock apart by the heat from the sun, by frost, by the roots of growing plants, by abrasion by other rocks, by the grinding action of ice, and by chemical change. Once the fragments are freed from their vaults, they are borne downward by wind, water, and ice, and the process of placer concentration is begun.

Some gold is concentrated without being moved. If a rock containing gold has been broken up, and some of the lighter material removed by wind or water, the deposit is called a *residual placer* (the rich *seam* diggings of the California Forty-Niners). As rough-gold bits released from a vein are tumbled about, or rolled along the bed of a stream, they soon are pounded into smooth *nuggets*. Therefore, if you find rough gold, you probably are exploring a residual deposit. Keep your eyes open: you may be very near the mother vein.

Though a stream or river may be born of torrents, or of huge, fast-melting snowfields, it is a thing of the moment.

From year to year, from storm to storm, from day to day—even from hour to hour—the stream bends, altering its course a little or a great deal. Along its banks one can see last year's high water mark, this year's flood plain, or the marks of a devastating torrent of several years ago.

All sizes of gold fragments are moved by the water of the rivers and streams. Though they may be carried along by the force of the stream some distance from their source, they sink with surprising speed toward the bottom of the water. *"It is almost impossible,"* wrote the late Dr. Waldemar Lindgren, foremost student of American ore deposits, *"to lose a particle of gold, of the value of one cent, in a miner's pan; it sinks immediately. . . .Once lodged at the bottom it stays there, in spite of shaking and rotating."*

This would lead one to think that all gold should be on the bottom of the stream bed. As everyone who has ever tried mining knows, that is not the case.

As the gold particles are moved downstream, they are in company with a large mass of slowly moving sand and gravel that slithers along the stream bed. Large nuggets may fall out along the way, being too heavy to be readily carried along by the water. They commonly stay nearer to the source than the medium and fine sizes, perhaps dropping in a hole in the lee of a large rock. Some of the minute gold particles—flour gold—are fairly easily sped down the rivers to the sea. Such tiny mites as this can be dissolved by the sea water, and become part of the sea itself. The sea actually contains as much as 65 milligrams of gold per ton of water.

The velocity and volume of a stream are greater in the center than at the edges, so that the amount of rock and gold a stream can carry is less on the sides than in the middle. A gold particle may be worked over to the stream's edge, to be suddenly dropped because the stream is no longer strong enough to push it along. It may stay there until it is found; or it may only stay until another freshet adds new vigor to the stream.

Swift water can carry much gold:

as its power lessens gold drops—

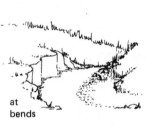
at bends

in still water

behind obstacles

on flood plains

The high mountains of the Sierra Nevada are on the skyline (Mt. Morrison to the left). Prospectors have wandered all through this rugged country in search of gold and other valuable minerals. There are gold, silver, and tungsten mines (and some with all three ores) scattered along these eastern foothills and high into the peaks. Rocks of the desert mountains, such as those in the foreground, have also been hosts for gold and silver deposits.

Desert country such as this is not a good place to dive for gold, as water is scarce, but one can employ dry placer methods, using the wind as a winnowing aid. Photograph by the author.

The "East Face" of the Sierra Nevada near Convict Lake.

Streams also lose their power when they change their course. As most streams bend and twist frequently, this means that there are many places where gold may be dropped. Usually this is on *bars* on the inside of bends.

As a stream turns, eddies and whirlpools are formed. If a gold nugget is suddenly thrown into the vortex of one of these whirlpools, it will be flung to the outer edge, there to join other heavy particles, and be gradually shaken downward toward bedrock. It is more likely that a nugget of gold will find its way into the eddies of the inside of the curve—the suction eddies—than to those on the outside, the pressure eddies.

As the gold fragments are pushed downstream and downward, they are sorted by nature into a group of heavy materials that, when suddenly dropped, will become a *pay streak* in a lens of sand and gravel.

Not all pay streaks and bars are to be found at present stream levels. In such rugged mountain regions as the Sierra Nevada of California, many of the streams have cut their channels deeper and deeper so that now their former flood plains or the bars at the bends of their former higher courses may remain, perched above the present level of the stream. In the early days of the California gold rush, miners found gold

in the present stream beds; those who were more canny also looked higher, searching for these perched beds, tracing them up the slopes until they found the grandfather bars more than half a mile higher. As the gold in the streams and on the high benches became more  difficult to win, the miners found still another source of placer gold, this time in bars of rivers that had flowed millions of years earlier.

But most recreational placer miners of today are not prepared to search for or mine these older deposits. They are mostly concerned with the gold in and along the rivers of today. And today's rivers still carry gold.

The bed of the stream itself is a very good place to look; for the heavy gold, as it is jigged along, falls to the bottom. There it is pushed along fairly rapidly over smooth rock (such as clean, unweathered granite or slippery serpentine), more slowly where the bottom is rough and uneven. It is wise to look where the bottom is ragged, for here the natural jigging action of the stream may concentrate gold in the lee of obstructions, which serve as natural riffles. If the bottom is soft and weathered, as in old schist or slate, the particles of gold may burrow deeply; where the bottom is markedly uneven, as in areas of deeply pitted limestone, particles of gold have been known to drop into crevices as much as 50 feet below the general stream bottom. In the early days, whole rivers were diverted from their courses in order to mine the stream bed; more recently a great deal of our American gold was won by huge dredges that mouthed great quantities of the river bottoms, spitting out boulders and debris and stacking them in neat but barren piles along the sides of the rivers.

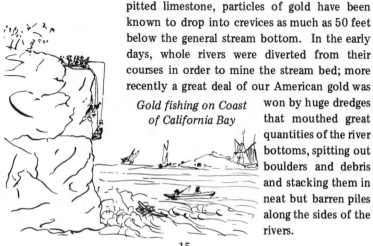

Gold fishing on Coast of California Bay

WHAT TO TAKE

Experience alone can tell a person what is the best outfit for him to take prospecting. The lists below will provide a starting point, but each miner will want to modify the equipment to suit his—or her—own needs. There are many books written on the subject—some dating back thousands of years. It is a wise idea to consult some of them well in advance.

For those who plan to skin dive for gold, consultation with experienced divers—perhaps at a sporting goods store specializing in underwater equipment—is advisable.

GENERAL

Lunch	First aid supplies	Fire permit
Maps	Matches in waterproof case	Compass
Waterproof flashlight	Gloves	Notebook and pencil

FOR DIVING

Air hose and supply	Fins	Canvas shoes
Safety rope	Face mask & mouthpiece	Life jacket
Suit, belt, weights	Underwater gloves	

FOR MINING

Gold pan	Tweezers	Prospector's pick
Knife	Broom, dustpan	Shovel
Sluicing equipment	Lightweight bucket	Magnet

Plastic sacks or bottles with firm lids for nuggets
Crevicing tools (spoons, pry bars, ice picks, etc.)
Sturdy sacks for carrying concentrates

Optional: an underwater gold sniffer

Above all else, if mining is to be a pleasure rather than a calamity, pay particular attention to highway, mountain, desert, and water safety.

Most prospectors will need, in addition to some version of the equipment listed on the opposite page, some basic lightweight camping equipment and transportation. A four-wheel drive vehicle is ideal for back roads and regions with no roads, but the casual prospector may not wish to invest in one. Whatever the vehicle, one should make certain that it is equipped for emergencies with such equipment as first-aid supplies, flares, and extra water and oil.

SOME

Do's

ON LAND

Do keep careful track of your location
Do carry food always
Do carry water if you leave a running stream
Do carry a flashlight and matches
Do wear a hat or shade in the summer sun
Do carry a simple first-aid outfit (snakebite kit included) and learn to use it
Do be exceedingly careful of fire
Do tell someone where you plan to go

IN THE WATER

Do learn water safety, particularly underwater safety
Do *learn and practice* mouth-to-mouth artificial respiration
Do keep your head

Don'ts

ON LAND

Don't go into the mountains alone; even the old-timers had burros
Don't become separated from your party
Don't destroy the landscape by thoughtless or careless actions
Don't leave litter

IN THE WATER

Don't dive alone. Someone should be out of the water, watching carefully at all times
Don't dive below 30 feet unless you are an expert
Don't hold your breath while ascending or surfacing
Don't get overtired in the water
Don't panic in tight places
Don't dive in exceedingly swift water

WHERE TO GO

Colorado's gold deposits

The best hunting ground for today's prospector is along streams in the gold country. Choose one that runs through a present or past mining area. Arm yourself with the tools that have been found most useful today. Take a gold pan to check for *colors*—small flecks of gold—in the stream gravels, a whiskbroom and dustpan, and a crowbar. After you have determined with the pan that there is some promise in the stream bed, walk along the water's edge until you find a rock that has cracks and crevices that have been under water during the winter. Pry the crevice open with the crowbar, then sweep out the cracks with the whiskbroom, being careful not to lose the fine clay. Pan this, or save it until you have enough to concentrate. After you have checked the crevices and cracks, look upstream of large rocks and tree snags, along the eddies and whirlpools. If you find black sand or clay, save it.

Those who have come prepared to go underwater should look in the same places: in cracks, crevices, behind obstructions, around roots of plants that grow in or near the water in the uneven bottom of the stream bed.

"All that glisters is not gold" is, unfortunately, quite true. Iron and copper pyrites both glitter and are moderately heavy, but not nearly so heavy as gold. Most fragments of them have flat, sometimes striated faces, rather than the lumpy appearance of gold nuggets. They are also quite brittle, and somewhat different in color. Mica, too, is quite glittery underwater, but the small flakes will float away during panning.

It is difficult to say exactly where one may go legally. Certainly on all open public land (this includes land under the control of the Bureau of Land Management), of which there are virtually millions of acres in the desert country of the western states, on all National Forest land (but not in National Parks and Monuments—except Death Valley—or in State Parks) and on navigable rivers. In general, except on posted private land, the gold-panning prospector will probably not be disturbed.

Staking a claim is quite another matter. It is wise to consult the state bureau of mines before attempting to do so.

Almost any of the rivers draining California's mother lode, or rising in the Klamath Mountains of northern California, are hopeful. The Salmon, the American, and the Yuba have been particularly fruitful. Colorado's rivers and streams are also good hunting ground, though it is wise to choose ones that rise in the mountains, rather than those on the plains, unless they have gold-bearing tributaries.

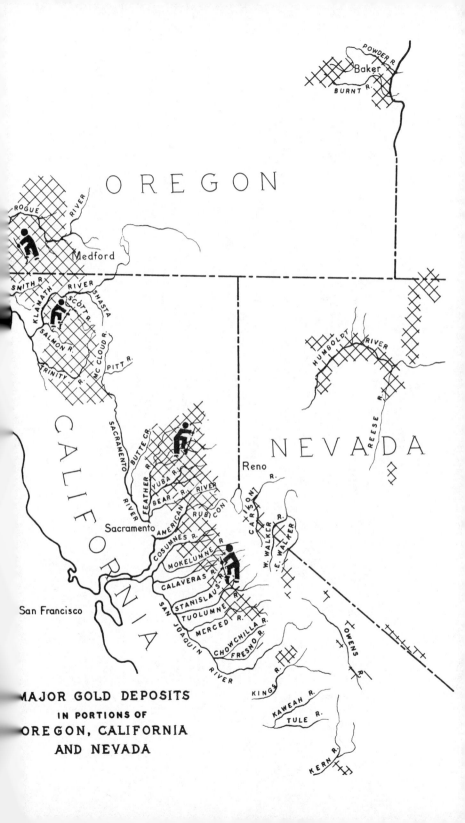

TIPS FROM

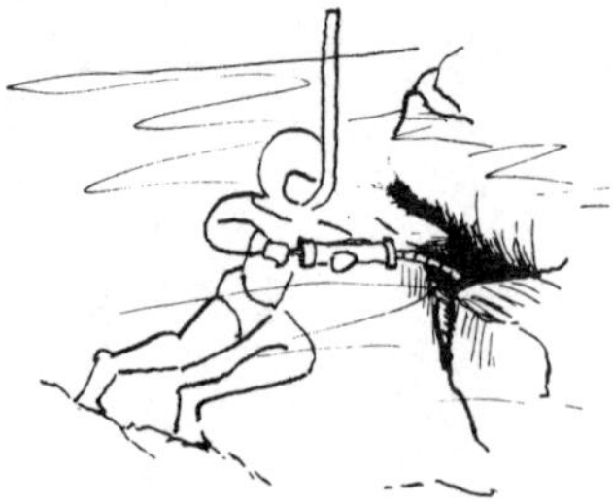

Clean out cracks and natural riffles

Check the junctions of creeks and rivers along the creek bank; if results are good try the creek aways.

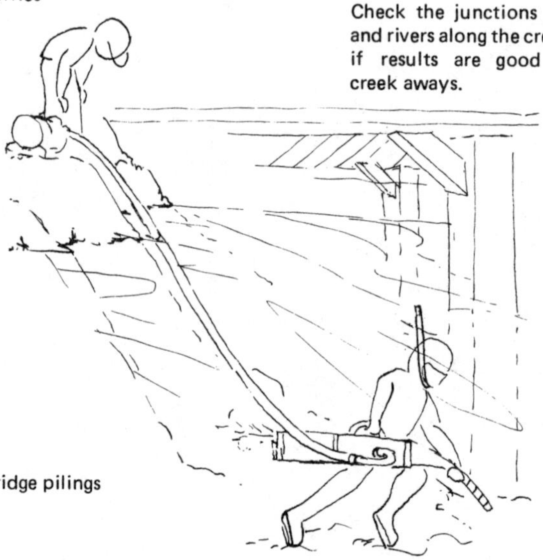

Search bridge pilings

Check around whirlpools rather than in their centers

Look below dams

OLD TIMERS

Sample down to bedrock; into it, if it is soft.

Puddle clay

Dig through "false bedrock" maybe the Chinese left something

Dig around boulders

Try the inside of bends

Look for eddies and swirl holes

Brush out crevices at the edge of the stream

Investigate stumps and snags

Don't worry with swift waterfalls

Gold in California, as in many other places in the world, is *"free gold"*—that is, it is not chemically tied to other elements. Because of this, anyone who can find gold can recover it without using complex machinery.

By dint of hard work, one miner can process one cubic yard of gravel per day by using a gold pan; two cubic yards can be processed in a rocker; a sluice box, which uses a great deal of water, can work 10 to 20 cubic yards of virgin gravel each day. When hydraulic mining was developed, using jets of water under high power, as much as 1,000 yards of gravel could be worked each day—1,000 times as much as a man with a pan! Until legal restrictions became so stringent that the mines could not operate, hydraulic methods could yield a profit from ore that carried as little as three cents of gold per ton!

Woodcut from *Underground* by Thomas Knox, published in 1873.

TOOLS

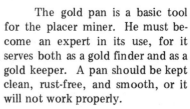

A California miner, 1842

The gold pan is a basic tool for the placer miner. He must become an expert in its use, for it serves both as a gold finder and as a gold keeper. A pan should be kept clean, rust-free, and smooth, or it will not work properly.

Commence by filling the pan about three-quarters full of gravel. Immerse the pan in water. Break fragments of dirt and clay, and throw away the larger rocks.

The principle upon which the gold pan is worked is the same that nature employs: the lighter material is washed away, the heavier—including the gold—left. A miner accomplishes this by moving the pan in a circle, thereby making a current, and combining the movement with a slight jerk to cast out the worthless pebbles and sand.

When there is left in the pan only black sand and gold, the panning is complete. Ordinarily this mixture of *concentrates* is kept until a quantity accumulates, then is amalgamated with mercury.

The pan also serves as a detecting tool. Use it to probe along the stream bed while looking for likely spots, panning here and there to check. One rule of thumb says: if there are seven flecks of gold in the first panful, try your luck there. You may have a pay streak.

The best way to learn to pan is to watch an experienced miner; he can quickly demonstrate the fundamental operation.

Though the gold pan is the most versatile tool for prospecting, for concentrating, and for detecting gold-bearing streaks, the miner may soon find he needs a piece of equipment that will do some of its work at a greater rate.

An experienced panner can process as much as a yard in 10 hours, but this is rarely a satisfactory amount. Sooner or later, some larger sort of gold-saving machine is wanted. Since the days of the forty-niners, thousands of machines have been built, from such simple devices as the rocker and the long tom to highly mechanized dredges capable of masticating whole river bottoms in a short time.

Basic to most of them is a series of *riffles*—baffles that are arranged to continue the work of sorting that the stream itself has started. They are usually mounted in a box of some sort, and are often staggered so that the water will run both around and over them. The lighter material, as in the gold pan, is washed away; the gold and heavy concentrates remain on the upper side, behind them.

The riffles and the box (together they form a *sluice box*) may be made of almost any material, though the lighter it is, the easier it will be to carry. Usually the box and riffles are made of wood, iron, or steel (though stones, poles, or other obstacles may be used), with napped cloth attached to the riffles to catch gold more effectively. The cloth is periodically washed and the wash-water panned to recover trapped gold.

A very simple sluice box is the *Indian log*, which is merely a set of notches cut into a log. The log is

RIFFLES

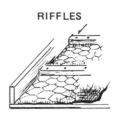

For fine sand cover with carpet and poultry cloth.

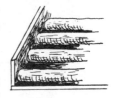

Logs set across—

or long way

Carpet or burlap can be put over or under any of these wood cleats.

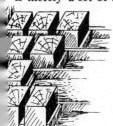

d grain blocks.

Cobble stones for rough work.

Angle bars with angles downstream.

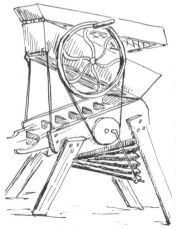

Dry washer

anchored in the stream in a likely place. In the early summer, it can be pulled up and the winter's gold recovered.

Many commercial sluices are on the market, including underwater boxes with enclosed riffles. Some are made of lightweight fiberglass or metal.

The sluice box or its equivalent should be arranged so that there is a minimum of labor required to operate it and a maximum of gold is saved. It should be placed so that a suitable quantity of water is easily available (preferably running into the box under its own power) and so that the operator does not have to shovel the material too high each time.

The head of the box should be set higher than the foot, so water will run through. Be sure to set the sluice where the *tailing* or waste material that has been washed through will not cover gravel and sand that is to be washed later. Most prospectors dig a *tail race*—a ditch for the waste water that will lead it away from the workings.

The angle of the sluice box is important. Too steep an angle will require a great amount of water, and may give sufficient power to the water that it can carry away the gold; too low an angle may make the operation too slow, or may form a sludge that keeps fine gold in suspension.

The "blanket toss", 1849

Placering in the Sixteenth Century

In the desert areas of the west, there are many *dry placers* that contain gold, but that can scarcely be worked by using water as the concentrating medium. Some of these are *bajadas* (bajada is the Spanish word for slope) that lie against the desert hills. Though there is no water near them during most of the year, they are truly water-formed deposits, that take their origin from torrential desert cloudbursts. Because the storms are infrequent, and, though mighty, are exceedingly brief, there is little time for the water to do detailed sorting. As a result, bajadas are leaner in gold than stream placers. North-facing slopes are usually richer than southern ones, as the water lingers there longer in the roots of the few desert plants.

Surprisingly enough, the wind itself is sometimes strong enough to move and sort gold. The white quartz dunes of Clayton Valley, Nevada, contain lenses of the precious metal—presumably torn from deposits in the surrounding hills.

To mine this dry-land gold, the wind must be made to do the work that water does elsewhere. Early Californians used the *blanket toss*, throwing the material high in the air so that the wind would winnow the chaff. In the Australian bush, miners became *dry-blowers*. The rich, gold-bearing desert dust was poured through the air from one pan to another, so that the wild *willy-willy* would blow away the worthless powder.

Rocking the cradle, 1849

Chinese miners were among the best in the California mines. Although often forced to work poorer ground, or areas already considered mined out by others, the Chinese patiently labored by hand, winning even the very finest gold dust.

Shown here is a rocker, one of the common gold recovering tools. It is operated by one or two men, one to cast out the larger, barren rock fragments, and to supply water, one to rock the machine back and forth, washing and winnowing the heavy gold.

One man with a small rocker can process twice as much gold as he could with a gold pan. Most miners use pans as prospecting tools, but turn to more efficient methods for actual gold recovery.

Woodcut from *Underground* by Thomas Knox, published in 1873.

A dry-placering machine may be constructed to utilize the wind by placing a series of screens one above the other. Though the wind blows hard in the American southwest, it bloweth when it listeth, and some more reliable machine is preferred by most dry miners. Such machines—*dry-washers*, as they are called—consist of a screen through which fine material drops onto a set of riffles. A bellows blows over the riffles—instead of the wind—gusting away the lighter dust. Illus. **page 26.**

Some gold, mostly in tiny, polished fragments, may travel to the sea. There, where the rush of water from the rivers is stilled, it is suddenly dropped. But the waves and currents and the tide continue to churn and sort it. It may gradually sink downward to find the lower reaches of

Diving for gold in 1915

the river delta, where, perhaps, at some distant date, the upheavings of mother earth may lift it from the water and make it a pay streak along the ocean shore. Many such fossil beaches, complete with pay streaks, have been found; the shorelines of other ancient, now dry seas, known to have contained gold, remain to be prospected.

Sometimes the sea itself reworks these once-immersed deposits. The beach gold and the gold from the modern rivers may be pushed by the longshore currents to form a thin band along the shore. Most of these—they are found in Alaska and along the Oregon and California coasts—contain their gold in black sand streaks. The beach at Nome, Alaska, though it is but 200 feet wide, has yielded more than 2 million dollars in gold.

The development of portable, inexpensive equipment for diving— so-called *skin diving*—has opened a magnificent underwater world that previously had been known to but few. The pleasure of living, for a time, in this totally strange environment is ample reason for skin diving.

The art of diving for gold is an old one, though most rarely practiced until the last decade. Very early gold seekers, who sought treasure below water as much as 2,000 years ago, used no equipment, and their methods could scarcely be called mining. After the development of deep-sea equipment, divers operating from dredges worked here and there in the gold-bearing rivers.

Though much of the easily won gold is gone now (particularly in such places as California, where careful Chinese miners scraped the pot quite clean in the early days) there is still some left; for during the last century streams and rivers have run through the gold veins to re-enrich the placers of yesteryear and provide wholly new ones unknown to the old-timers.

A "surf washer" for gold

"Skin diving", as one diving geologist wrote, *"is deceptively easy. You can learn enough in five minutes to kill yourself."* Surprisingly, the main danger, particularly for prospectors, is the danger of drowning.

A diver—who should be an expert swimmer—should be exceedingly careful not to tangle himself in his own line; he should not go below 30 feet (rarely necessary for placer mining) without decompressing himself as he surfaces; he should never hold his breath while ascending, as he may rupture blood vessels within his lungs. In some areas, there are skin diving schools in which safety and techniques are taught. Almost everywhere, a more experienced diver can be found to consult.

Even in the far north there have been skin-diving miners who won gold from the bitter depths of the Fraser River. Clad in *ice-water suits*, these intrepid argonauts have chopped through several feet of ice in weather of $15°$ below zero and worse, to dive for—and find—gold. In

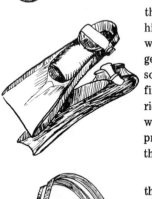

the tropics, though miners have dived for gold and diamonds, there are still long miles of alligator-guarded rivers that may be very rich, as yet untapped by skin divers.

Basic equipment for the underwater gold prospector are his mining tools and his diving gear.

Many of his mining tools will be the same as those used by prospectors on land; indeed, much of his work will be done at the riverside. His diving gear will, on the other hand, be totally different from any gear carried by dry-land prospectors. He will need some air supply with hoses, regulator, a face mask, fins, and weights. As gold prospecting is usually carried on in cold mountain streams, most divers wear a wet suit; in rocky areas, they may wear coveralls to protect their fragile suits, and canvas shoes to protect their feet.

Experienced prospector-divers have found that the proper amount of weight to be worn is just enough to stay on the bottom, an amount that varies with the diver's weight. Most of the weight is carried on the weight belt, though weights are also added to the feet in order to stand upright. Small nugget bags, lights, crevicing tools, and other useful equipment can also be hung on the weight belt, leaving the hands free to move large rocks or to clean out cracks.

Underwater breathing equipment is of several types, ranging from none (hold your breath and go) to complex deep-sea mechanisms. In general, there are three types—*SCUBA* (self-contained underwater breathing apparatus), *OBA* (oxygen breathing apparatus), and *HOOKAH* (hose) diving, Scuba, consisting of a mask, tube, and tank carried on the diver's back, and hookah diving, are the two best suited to prospecting.

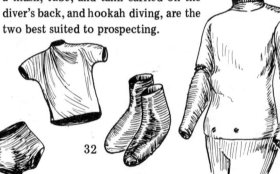

For hookah diving a suitable compressor unit (hoses and unit together are called *hookah)* must be located nearby, on shore or on a pontoon raft or tip-proof boat. Diaphragm compressors are used to maintain purity of the diver's air. At all times there should be some person above water keeping careful track of the diver, whether he is also responsible for the diver's air supply or not.

Underwater dredges that run on compressed air are favorite underwater mining equipment. They consist of suction tubes that collect material in the manner of a vacuum cleaner. Some of those on the market send all the mud, sand, gravel, and gold through the impeller; others, better for mining purposes, do not. Some also have underwater sluices attached to or built into them; with others, it is necessary to sluice the material on land.

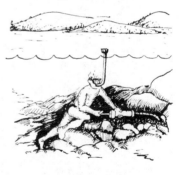

A simple suction tube called a *gold sniffer*—actually a large syringe—is a poor man's adaptation of a suction dredge. It can be used in addition to the dredge, as well, for it will clear out small corners the larger dredge cannot enter. Such a *sniffer* can be made by converting a grease gun, though there are commercial ones available.

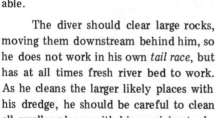

The diver should clear large rocks, moving them downstream behind him, so he does not work in his own *tail race,* but has at all times fresh river bed to work. As he cleans the larger likely places with his dredge, he should be careful to clean all smaller places with his crevicing tools.

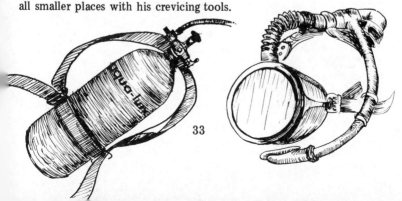

AMALGAMS AND RETORTS

Cut a well rounded white potato in two and scoop out a space in one half.

Pack amalgam in space and wire potato together.

Wrap in aluminum foil and bake in hot ashes for an hour.

The potato will absorb the mercury and leave a gold button.

Mercury and fumes are POISONOUS

After the sluice box has yielded its black sand, and after the prospector's pan has separated out all the worthless light rock and sand, there usually remains a black sand that contains the gold together with other heavy minerals. Titanium compounds, magnetite, chromite, garnet, hematite, platinum—all may remain in this black mass.

As much of it is not particularly valuable (at least not in small quantities), the gold is usually separated and saved, the rest cast away. Some prospectors can do this by using their thumbs dextrously; most, however, rely on mercury to amalgamate the gold.

The amalgamation process can be carried out in a pan, but it should not be the same pan that is used for prospecting. It is nearly impossible to avoid leaving a thin film of mercury on the metal surface, which may mask the gold *colors*—fine gold flecks—so that they cannot be seen in the next panning. If, for some reason, the prospecting pan must be used, it should subsequently be scraped and burned.

A few drops of mercury are placed in the pan, then rolled about. They race among the gold particles, gathering them into the body of the mercury (quicksilver) drops.

Some prospectors prefer to use copper-bottomed pans in which the copper itself is amalgamated. This is accomplished by placing a bit of mercury on the copper bottom, then rubbing it "into" the copper, as children love to change copper pennies to "silver" ones with a bit of mercury. A bottom thus treated will, when a few drops of free mercury are added, cling tenaciously to any gold that drops on it. The mercury and gold amalgam cannot be poured out, as it can from a steel or iron pan, but must be scraped gently.

Sometimes mercury is used in the riffles of a sluice box while the water is running through it.

It makes a magnificent trap for gold, but one must be careful to not let the mercury wash away under too strong a head of water.

It is possible to retort the mercury to separate it from the gold, and to use it again. However, if a small amount of mercury is involved, it may be just as well to allow it to volatilize into the air, without trying to recover it.

There are several ways of accomplishing the dissociation of gold from mercury. One fairly simple way is to cook it on a pan over a low fire. Be sure to do this in the open air; at one time an entire family died from attempting this process over the home stove.

If the mercury is still wettish, place it in a cloth sack (sugar or salt sacks are good for this purpose), and squeeze. The liquid mercury, freed from gold, will pour out, leaving a comparatively dry residue to be retorted.

On these pages are two methods of retorting amalgam. Using the potato, the mercury is lost (so is the potato!—don't eat it!); in the home-made retort designed by Messrs. John Gow and John Jan, the mercury can be saved for re-use.

Use black steel pipe and fittings—have machine shop drill and tap a 2" pipe plug for ¼" nipple.

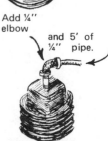

Add ¼" elbow and 5' of ¼" pipe.

You will also need a 2" pipe cap.

Coat all the threads with chalk or clay before assembling.

Wrap one ounce of amalgam in a little paper and put it in the pipe cap. Screw the plug assembly and cap with amalgam together. Use clay or chalk on these threads also.

THIS IS FOR OUTDOORS ONLY!

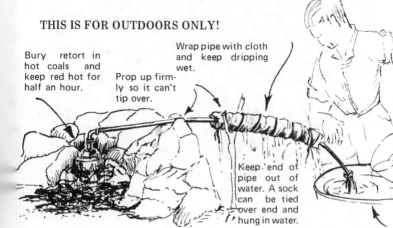

Bury retort in hot coals and keep red hot for half an hour.

Prop up firmly so it can't tip over.

Wrap pipe with cloth and keep dripping wet.

Keep end of pipe out of water. A sock can be tied over end and hung in water.

Tub or bucket of water.

LIFE AND DEATH OF A MINE

One of the oldest of California's lode gold mines, the Kennedy was discovered in 1856, and soon became one of the most important in the mother lode. The main buildings of the mine are shown in this old photo, taken about the turn of the century. In the background, Jackson Butte may be seen, a few miles to the south. Today, the modern town of Jackson has grown northward nearly to the old mine.

Underground, miners of three-quarters of a century ago pushed ore cars through 150 miles of workings, illuminated by candles. Hard safety hats had not yet been invented; miners wore soft caps—scant protection against falling rock. For many years, the Kennedy held title as the deepest mine in

North America, as it burrowed more than a mile below the surface of the ground.

A great deal of gold—more than sixty million dollars worth at today's prices—was won from the old mine in its lifetime, but the mine had its tragedies as well as its successes. In 1922, a whole shift of miners at the adjoining Argonaut (the mines were continuous underground) was trapped by a fire. Eventually, rescuers worked through the Kennedy to them, but 47 were already dead. The photo on page 40 shows the message left by the dying miners.

The mine had its ecological problems, too. To send the poisonous waste from the mill away from the water supply to a special impounding dam, the huge wheels shown in the photograph on page 41 were constructed. There were four of them altogether, each fitted with 176 little buckets to lift the waste—tailing— a vertical height of 48 feet, so that the slushy material could be eased downhill into a catchment basin. The wheels themselves were each 68 feet high, made of pine and redwood.

A small park now marks the site of the four

wheels. One has been reconstructed; the other three are in various stages of disrepair.

The wheels last ran in 1942, when, by presidential wartime emergency order, the mine, together with all mines in the United States that produced only precious metals, was shut down. To be sure, there is still much gold in the now-flooded, partly caved mine workings, but the cost of rehabilitating the mine, the high price of labor, as well as the lack of experienced miners, has so far made the cost of reopening it impractical.

Photographs on pages 36-37, 38-39, courtesy of California Division of Mines and Geology, photographer unknown. The wheels photograph opposite is by the author.

Note left at Kennedy disaster; illustration is from U.S. Bureau of Mines Technical Paper 363, page 15.

WHERE TO SELL YOUR GOLD

Until recent years, it was illegal in the United States to sell gold to anyone except the U.S. Government or a licensed gold buyer. Currently, the U.S. Mint does not buy newly mined gold, and one may sell gold to any buyer.

The price you may receive for gold is no longer fixed by law at $35 per ounce, as it was for nearly forty years. Today, you will receive a price close to the current market value, modified by the state of your gold. Most gold purchasers will not buy black sand, crude ore, or low-grade concentrate. You must furnish them with gold grains or gold nuggets, gold sponge resulting from amalgamation with mercury, or crude gold bullion.

Once you have refined your gold to an acceptable condition, ask your state geological survey or one of the U.S. Geological Survey Public Inquiries Offices for information on the nearest possible buyer. If you live near a large gold mine, perhaps that company will buy it; the Homestake Mining Company, P.O. Box 875, Lead, South Dakota 57754, has, in the past, bought gold from independent miners. Their mine is the largest in the nation.

Homestake Mining Company also operates a gold refinery, so that they are able to accept scrap gold and gold-bearing alloys, as well as reasonably pure sponge gold and high quality placer gold. Other gold refineries may also accept similar gold. Here are the addresses of some:

American Smelting and Refining Company
P.O. Box 151
Perth Amboy, New Jersey 08861

Western Alloy Refining Company
366 East 58th Street
Los Angeles, California 90011

Wildberg Brothers Smelting and Refining Company
349 Oyster Point Boulevard
South San Francisco, California 94080

Eastern Smelting and Refining Corporation
37-39 Bubier Terrace
Lynn, Massachusetts 01901

Handy and Harman
4140 Gibson Road
El Monte, California 91731

If your gold is in large nuggets, or is unusual for some other reason, it may be worth far more than the market value as a specimen or collector's piece. Banks and museums as well as mineral collectors may purchase unusual specimens.

Photograph by Donna Jeffries.

WHERE TO GET MORE INFORMATION

The books and articles that have been written on gold are legion. Some of them are as old as the history of the world; some are the most exciting adventure stories that have ever been written.

The work listed below is a classic that has recently been revised and rewritten. It will not only serve as a very valuable field partner, but will also lead you to other useful information:

Handbook for Prospectors. Fifth edition 1973. By Richard M. Pearl, McGraw-Hill Book Company, New York, 472 pages. A revision of *Handbook for Prospectors and Operators of Small Mines*, by M.W. Von Bernewitz.

For those who are interested in desert placer mining—barely treated by most writers—there is this engaging article by that late grand old man of letters and science, T.A. Rickard: *"The alluvial deposits of Western Australia,"* in Transactions of the American Institute of Mining Engineers, volume 28, p. 490-537, 1898. Out of print, but available at some libraries.

Virtually every western mountain state except Hawaii has a publication on placer mining. Many of them are out of print, but may be consulted at libraries. The U.S. Geological Survey has issued a series of bulletins on placer deposits in the various western states; write the U.S. Geological Survey, Arlington, Virginia, 22202, for a free list of publications.

PUBLIC OFFICES OF THE U.S. GEOLOGICAL SURVEY

For mail order:

U.S. Geological Survey
1200 South Eads
Arlington, Virginia 22202

U.S. Geological Survey
Denver Federal Center
Building 41
Denver, Colorado 80225

Residents of Alaska only may order from:

U.S. Geological Survey
Distribution Center
310 First Avenue
Fairbanks, Alaska 99701

PUBLIC INQUIRIES OFFICES

U.S. Geological Survey
108 Skyline Building
508 2nd Avenue
Anchorage, Alaska 99501

U.S. Geological Survey
7638 Federal Building
300 North Los Angeles Street
Los Angeles, California 90012

U.S. Geological Survey
504 Custom House
555 Battery Street
San Francisco, California 94111

U.S. Geological Survey
1012 Federal Building
1961 Stout Street
Denver, Colorado 80202

U.S. Geological Survey
8102 Federal Building
125 South State Street
Salt Lake City, Utah 84111

U.S. Geological Survey
678 U.S. Court House
West 920 Riverside Avenue
Spokane, Washington 99201

STATE GEOLOGICAL SURVEYS OF WESTERN U.S.A.

Alaska Division of Geological Survey
3001 Porcupine Drive
Anchorage, Alaska 99504

Arizona Bureau of Mines
University of Arizona
Tucson, Arizona 85721

California Division of Mines and Geology
1416 9th Street, Room 1341
Sacramento, California 95814

Colorado Geological Survey
1845 Sherman Street, Room 254
Denver, Colorado 80203

Hawaii Division of Water and Land Development
P.O. Box 373
Honolulu, Hawaii 96809

Idaho Bureau of Mines and Geology
University of Idaho
Moscow, Idaho 83843

Montana Bureau of Mines and Geology
College of Mineral Science and Technology
Butte, Montana 59701

Nevada Bureau of Mines and Geology
University of Nevada
Reno, Nevada 89507

New Mexico Bureau of Mines and Mineral Resources
Campus Station
Socorro, New Mexico 87801

North Dakota Geological Survey
University Station
Grand Forks, North Dakota 58201

Oregon Department of Geology and Mineral Industries
1069 State Office Building
Portland, Oregon 97201

South Dakota Geological Survey
Science Center
University of South Dakota
Vermillion, South Dakota 57069

Utah Geological and Mineralogical Survey
University of Utah
Salt Lake City, Utah 84112

Washington Division of Mines and Geology
P.O. Box 168
Olympia, Washington 98501

Geological Survey of Wyoming
P.O. Box 3008, University Station
Laramie, Wyoming 82070

GEOLOGICAL SURVEYS IN CANADA

Geological Survey of Canada
Ottawa, Canada

Research Council of Alberta
87th Avenue and 114th Street
Edmonton, Alberta, Canada

Department of Mines and Petroleum Resources
Victoria, British Columbia, Canada

Department of Mines and Natural Resources
901 Norquay Building
401 York Avenue
Winnipeg 1, Manitoba, Canada

Department of Lands and Mines
P.O. Box 758
Fredericton, New Brunswick, Canada

Department of Mines and Resources
St. John's, Newfoundland, Canada

Department of Mines
Province of Nova Scotia
Halifax, Nova Scotia, Canada

Ontario Department of Mines
Parliament Buildings
Toronto 2, Ontario, Canada

Department of Natural Resources
Quebec, Province of Quebec, Canada

Department of Mineral Resources
Government Administration Building
Regina, Saskatchewan, Canada